Leaping Learners
Education, LLC

For more information and resources visit us at:
www.leapinglearnersed.com

Photo Credits

Monkey graphics © dreamblack46/stock.adobe.com

Cover, © Sergey Uryadnikov/shutterstock.com; Pg. 1-2 © Onyx9/shutterstock.com; Pg. 3-4 © max_776/stock.adobe.com; Pg. 5-6 © GUDKOV ANDREY/shutterstock.com; Pg. 7-8 © Kiki Dohmeier/shutterstock.com; Pg. 9-10 © N. F. Photography/shutterstock.com; Pg. 11-12 © dossyl/shutterstock.com; Pg. 13 © Eric Gevaert/shutterstock.com; Pg. 14 © bimserd/shutterstock.com; Pg. 15-16 © Patrick Rolands/shutterstock.com; Pg. 17 ©Kiki Dohmeie/shutterstock.com; Pg. 18 © Styve Reineck/shutterstock.com; Pg. 19-20 © Simon Eeman/shutterstock.com; Pg. 21-22 © KiltedArab/shutterstock.com; 23-24 © Rich Carey/shutterstock.com; Pg. 25 © nikolae/shutterstock.com, © Mari C/shutterstock.com

*All design by Sean Bulger*
*All other pictures by Sean Bulger or royalty free from Pixabay.com*

ISBN
978-1-948569-20-0

Dear Parents and Guardians,

Thank you for purchasing a *Matt Learns About* series book! After teaching students from kindergarten to second grade for more than seven years, I became frustrated by the lack of engaging books my students could read independently. To help my students engage with nonfiction topics, my wife and I decided to write nonfiction books for children. We hope to inspire young children to learn about the natural world.

Here at Leaping Learners Education, LLC, we have three main goals:

1. Spark young readers' curiosity about the natural world
2. Develop critical independent reading skills at an early age
3. Develop reading comprehension skills before and after reading

We hope your child enjoys learning with Matt. If you or your children are interested in a topic we have not written about yet, send us an email with your topic, and maybe your book will be next!

Thank you,

Sean Bulger, Ed.M

www.leapinglearnersed.com

# Reading Suggestions:

Before reading this book, encourage your children to do a "picture walk," where they skim through the book and look at the pictures to help them think about what they already know about the topic. Thinking about what they already know helps children get excited about learning more facts and begin reading with some confidence.

Preview any new vocabulary words with your child. Key vocabulary words are found on the last few pages of the book. Have your children use each new phrase in their own words to see if they understand the definition.

After previewing the book, encourage your child to read the book independently more than once. After they have read it, ask them specific questions related to the information in the book. Encourage them to go back and reread the relevant section in the book to retrieve the answer in case they forgot the facts.

Finally, see if your child can complete the reading comprehension exercises at the end of the book to strengthen their understanding of the topic!

This book is best for ages 6-8
but. . .
Please be mindful that reading levels are a guide and vary depending on a child's skills and needs.

# Matt Learns About . . . Tarantulas

Written by Sean and Anicia Bulger

# Table of Contents

Hi! My name is Matt. I love to discover and learn new things. In this book, we will learn about tarantulas. Let's go!

# Introduction

Look!

A creepy-crawly tarantula!

Tricky word:
say...
ta-ran-tu-la

# Habitat

Where do tarantulas live?

Tarantulas live in forests all around the world.

NORTH AMER

EUROPE

ASIA

AFRICA

SOUTH
AMERICA

AUSTRALIA

**Key**

Tarantula's habitat

ANTARCTICA

PG 4

Most tarantulas live on the forest floor. Some tarantulas live in trees.

PG 6

Most tarantulas live in **burrows** underground.

They make the burrow or they move into one that was already made by another animal.

Burrow

PG 8

Tarantulas make their burrow using their long legs and **fangs** to move the dirt.

PG 10

# Body

What do tarantulas look like?

Tarantulas have a large, hairy body with eight legs and two **pedipalps**. On their backside, tarantulas have an **abdomen**.

Pedipalps

Abdomen

Tricky Word:
say...
Ped-i-palp-s
Ab-doe-men

PG 12

Tarantulas have large fangs. They use these fangs to bite and inject **venom** into their **prey**.

Fang

Tarantulas have eight eyes, but they do not use their eyes to hunt for food.

Instead, they use hairs on the front of their legs. These hairs **sense** where a tarantula's prey is.

When a tarantula grows, it changes its old skin for new skin. This is called **molting**.

PG 18

# Food

Tarantulas love to eat insects like crickets and grasshoppers.

What do tarantulas eat?

Some tarantulas even eat mice, birds, or other small animals.

# Predators

Tarantulas are eaten by hawks, snakes, and spider wasps.

Which animals eat tarantulas?

# Staying Safe

How do tarantulas stay safe?

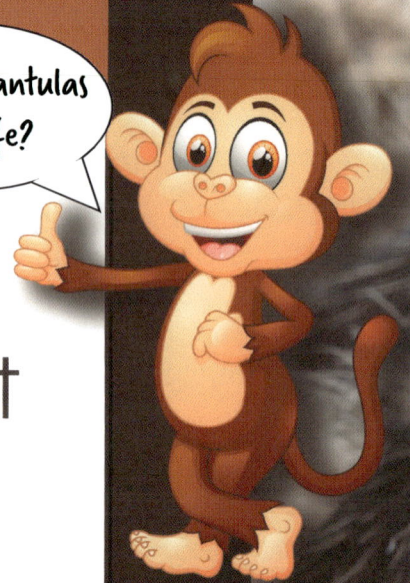

To stay safe, tarantulas can shoot sharp hairs off their abdomen at an enemy. This hurts the enemy and helps the tarantula escape.

PG 23

Hair

# Baby Tarantulas

Female tarantulas lay 50 to 2,000 eggs at a time. She carries the eggs with her until they hatch. After the eggs hatch, the baby tarantulas take care of themselves.

PG 26

What are some interesting facts about tarantulas?

!!!!!

A tarantula's mouth is like a straw.

A tarantula cannot kill a person.

Tarantulas are the biggest spiders in the world.

PG 27

# Glossary

A glossary tells the reader the meaning of important words.

**Burrows** – Holes in the ground made by animals

**Fangs** – Sharp and pointy teeth that a snake uses to catch prey

**Pedipalps** – Front appendage spiders use to taste and smell

**Abdomen** – Stomach of the spider

**Venom** – Poisonous liquid into other animals to kill or paralyze them

**Prey** – Food for other animals

**Sense** – To know when something is close by

**Molting** – When an animal replaces its old skin with new skin

Draw a picture of a tarantula.

Draw a picture of a tarantula eating.

The tarantula is eating a

_____.

# Quiz

**1. Where do tarantulas live?**

a. High in trees

b. In underground burrows

c. In rivers

**2. What is one word you can use to describe a tarantula's body?**

a. Hairy

b. Smooth

c. Bumpy

**3. What is molting?**

a. When a tarantula sheds its skin

b. When a tarantula makes a home

c. When a tarantula captures prey

Common core standards:
RI. 1. 1 - Questions 1, 2
RI. 1. 2 - Question 3

**4. What is one thing tarantulas eat?**

a. Grasshoppers

b. Fruit

c. Leaves

**5. What is venom?**

a. Liquid poison

b. Teeth

c. Hair

**6. What does the picture on page 12 teach you?**

a. What a tarantula's body looks like

b. How tarantulas eat

c. What baby tarantulas look like

Common core standards:
RI. 1. 1 - Questions 4, 5
RI. 1.8 - Question 6

# Want to learn about ocean animals? Check out the "Fay Learns About…" series!

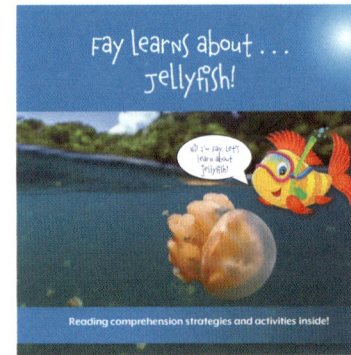

Great for emerging readers ages 6-8

# Want to learn about Farm Animals? Check out the "Katie Teaches you About..." series!

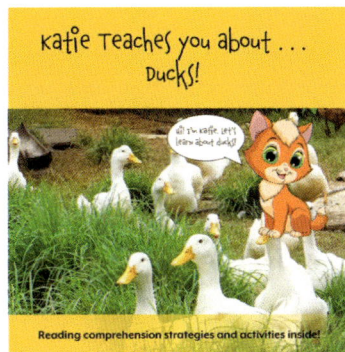

Katie Teaches you about . . . Ducks!

Reading comprehension strategies and activities inside!

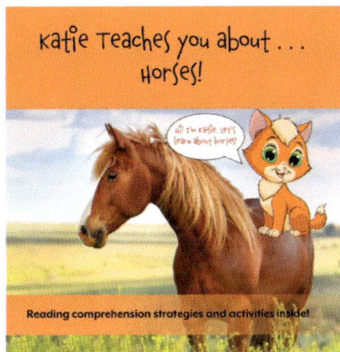

Katie Teaches you about . . . Horses!

Reading comprehension strategies and activities inside!

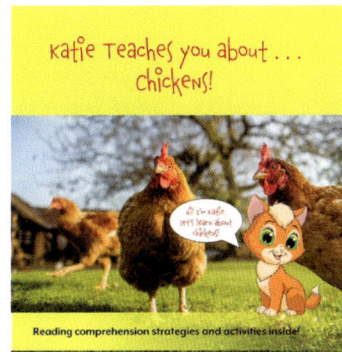

Katie Teaches you about . . . Chickens!

Reading comprehension strategies and activities inside!

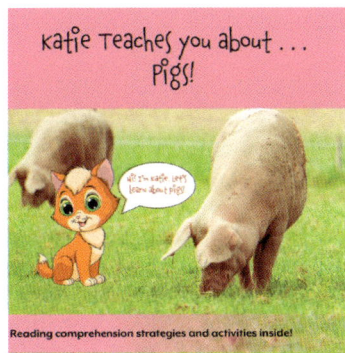

Katie Teaches you about . . . Pigs!

Reading comprehension strategies and activities inside!

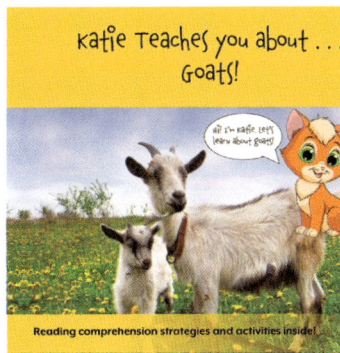

Katie Teaches you about . . . Goats!

Reading comprehension strategies and activities inside!

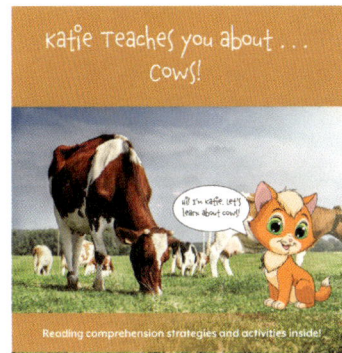

Katie Teaches you about . . . Cows!

Reading comprehension strategies and activities inside!

# Great for early readers ages 4-6

# Want to learn about colors? Check out the "Clayton Teaches you About..." Series!

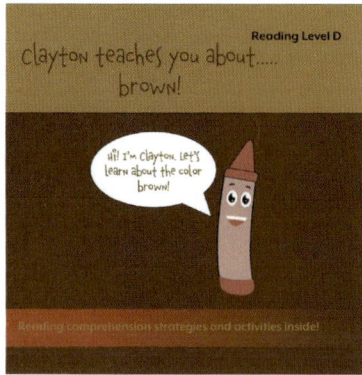

Reading Level D

Clayton teaches you about.....
brown!

Hi! I'm Clayton. Let's learn about the color brown!

Reading comprehension strategies and activities inside!

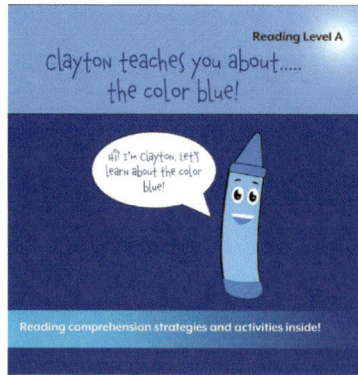

Reading Level A

Clayton teaches you about.....
the color blue!

Hi! I'm Clayton. Let's learn about the color blue!

Reading comprehension strategies and activities inside!

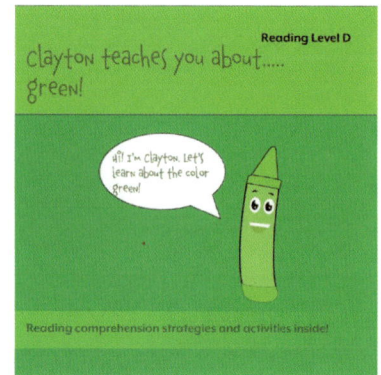

Reading Level D

Clayton teaches you about.....
green!

Hi! I'm Clayton. Let's learn about the color green!

Reading comprehension strategies and activities inside!

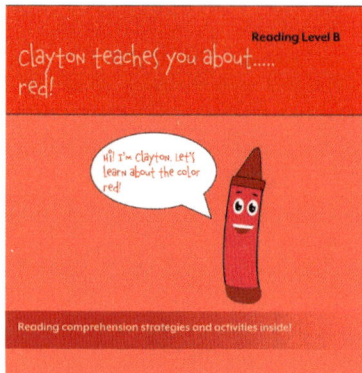

Reading Level B

Clayton teaches you about.....
red!

Hi! I'm Clayton. Let's learn about the color red!

Reading comprehension strategies and activities inside!

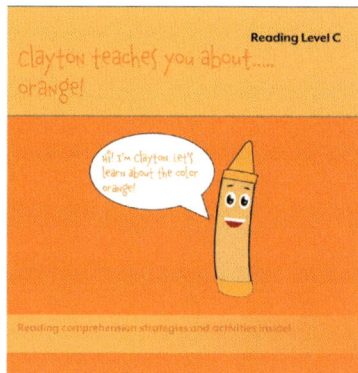

Reading Level C

Clayton teaches you about.....
orange!

Hi! I'm Clayton. Let's learn about the color orange!

Reading comprehension strategies and activities inside!

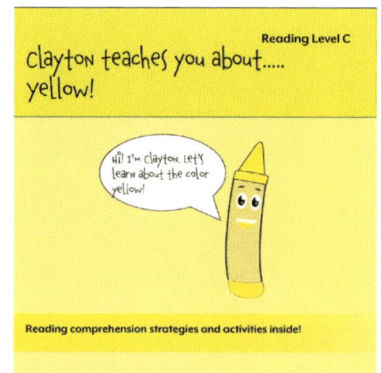

Reading Level C

Clayton teaches you about.....
yellow!

Hi! I'm Clayton. Let's learn about the color yellow!

Reading comprehension strategies and activities inside!

# Great for early readers ages 4-6

Printed in Great Britain
by Amazon